P9-DDV-789

WITHDRAWN

First Science
Weather Watch!

Editorial planning: Serpentine Editorial
Scientific consultant: Dr. J.J.M. Rowe

Designed by The R & B Partnership
Illustrator: David Anstey
Photographer: Peter Millard

Additional photographs:
Chris Fairclough Colour Library 6, 11 (top), 19, 24 (top);
The Hutchison Library 7;
Stephen Dalton/NHPA 9 (top);
ZEFA 10, 11 (bottom), 15 (top and bottom), 16;
John Shaw/Bruce Coleman 14;
NASA/Bruce Coleman 30 (top);
Bryan & Cherry Alexander/NHPA 30 (bottom).

Library of Congress Cataloging-in-Publication Data

Rowe, Julian.
Weather watch! / by Julian Rowe and Molly Perham.
p. cm. — (First science)
Includes index.
ISBN 0-516-08142-X
1. Weather — Juvenile literature. 2. Weather — Experiments — Juvenile literature. 3. Climatology — Juvenile literature. 4. Weather forecasting — Juvenile literature. [1. Weather. 2. Weather — Experiments. 3. Climatology. 4. Weather forecasting. 5. Experiments.] I. Perham, Molly. II. Title. III. Series: First science (Chicago, Ill.)
QC981.3.R67 1994
551.5—dc20

94-16944
CIP
AC

1994 Childrens Press® Edition

2 3 4 5 6 7 8 9 10 R 0.3 02 01 00 99 98 97 96 95

Weather Watch!

Julian Rowe
and Molly Perham

CHILDRENS PRESS®
CHICAGO

Contents

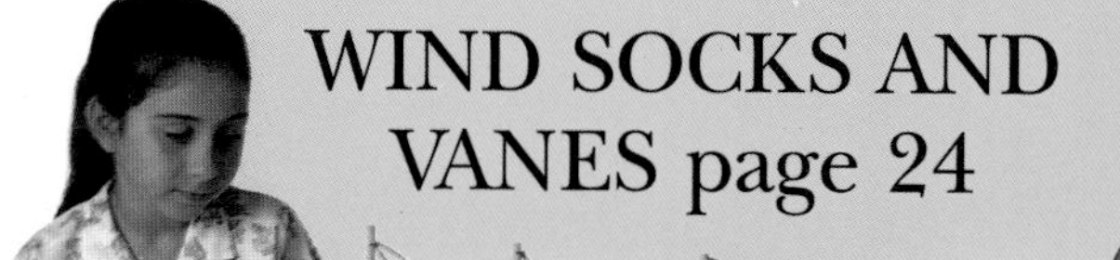

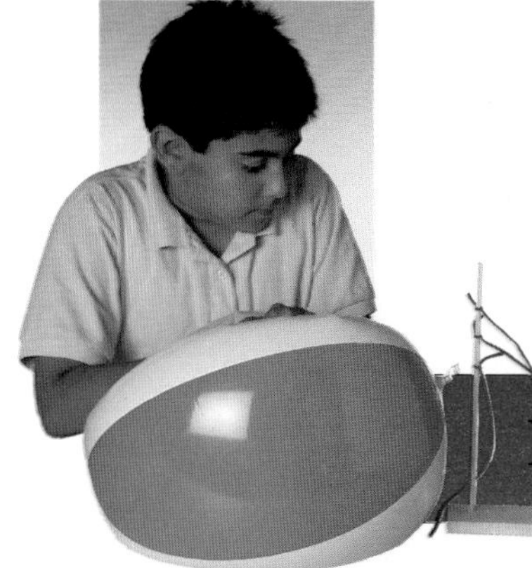

SAFETY WARNING

Activities marked with this symbol require the presence and help of an adult. Plastic should always be used instead of glass.

Hot weather

What is the weather like today?
Is it sunny and hot?
On hot days we wear light clothing
to help keep us cool.

These Arab boys live in the Sahara, where it is very hot. They wear long, loose robes and headdresses to keep off the desert sun.

The sun's rays can be dangerous if you let them burn your skin.

Cold weather

On cold days, we need plenty of clothes
to keep us warm.
We wear woolly scarves and gloves.

In winter, birds ruffle their feathers. This traps air around their bodies and helps keep them warm.

In cold weather, it is better to wear several thin layers rather than one layer of thick clothing. Air is trapped between the layers and keeps heat in.

The animal and plant world

Polar bears live in the Arctic, one of the coldest places in the world. They have long, thick fur to keep them warm.

Cacti store water in their fleshy stems. The prickly spines protect the plant and its precious water from desert animals.

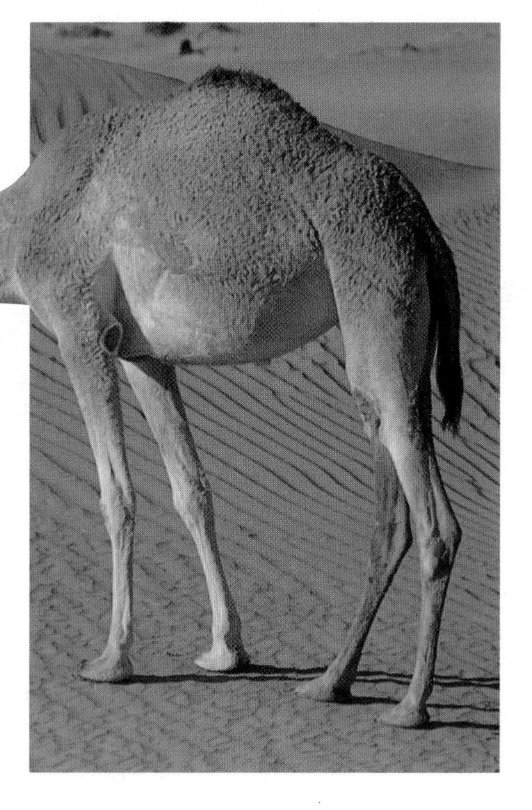

A camel is well-suited to life in the hot, dry desert. It can go for days without eating and drinking. It lives off fat stored in its hump.

It's raining!

When it rains, we wear waterproof clothing. Clouds form when water in the air, called vapor, cools and turns into tiny droplets. When the droplets group together they become so heavy that they fall as rain.

Scissors

Make a rain gauge

Materials: A plastic bottle, scissors, bricks, and a ruler.

1. Ask an adult to cut the top half off the bottle.

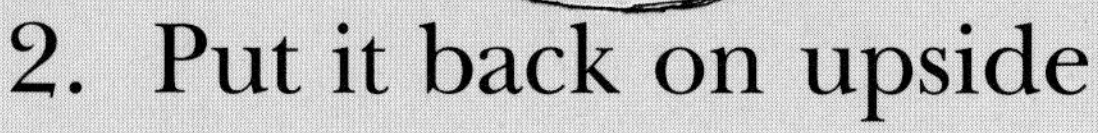

2. Put it back on upside down.

3. Stand your rain gauge outside. Put bricks around it to hold it steady.

4. Each day, measure how much rain has fallen.

It's snowing!

When it is cold, the drops of water in a cloud freeze into ice crystals and fall as snow.

Each snowflake has a different shape, but they all have six sides or points.

A heavy snowfall can cause problems by blocking roads and airports.

But what fun it is to slide down a slope on a sled!

Have you ever made a snowman? How long did it last?

It's freezing!

Snow melts when the temperature rises and it becomes warmer. Icicles form when the melted snow freezes again at night.
When water freezes on sidewalks and roads, it makes them dangerous to walk and drive on.

Investigating ice

Materials: Ice cubes, a plastic jug or bowl, and a thermometer.

1. Put the ice cubes in the jug.

2. Stand the thermometer in the jug and leave it for a few minutes.

3. What is the temperature on the thermometer?

4. How long does it take the ice cubes to melt?

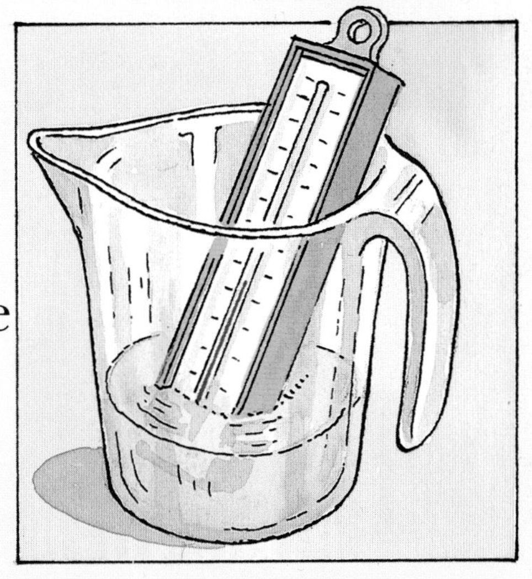

How much water do you think you would get from an icicle? How much from a bucket of snow?

Fog and mist

The air you breathe out contains water vapor.

When you breathe out on a cold day, the vapor changes back into tiny drops of water and a misty cloud forms. This is called condensation.

Fog and mist are formed in the same way.

They are made up of millions of tiny drops of water, like those in a cloud, but they are close to the ground. When it is foggy we cannot see very far.

Drivers must reduce their speed and turn on their lights.

Blowing in the wind

When you hang out laundry, water in the wet clothes turns into vapor, or evaporates.

On a windy day, the clothes dry faster because the wind blows the water vapor away.

The wind can blow your hat off your head.

Sometimes wind blows hard enough to turn an umbrella inside out!

Measuring the wind

The speed of the wind is measured with an instrument called an anemometer.

When a really strong wind is blowing, it is difficult to walk into it.

Make an anemometer

Materials: Three paper cups, three knitting needles, a large cork with a hole in it, a broom handle, a hammer, a long nail, and two washers.

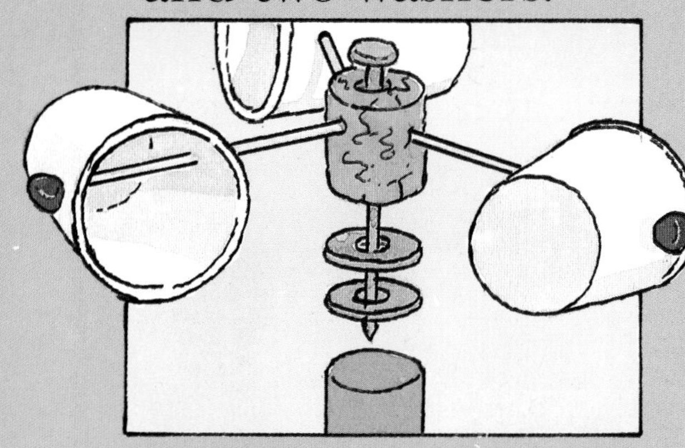

1. Push a knitting needle through a cup and into the side of the cork. Do the same with the other two cups. Push the nail through the hole in the cork and the washers.

2. Ask an adult to hammer the nail into the top of the broom handle.

3. When the wind blows, can you measure how fast the anemometer turns?

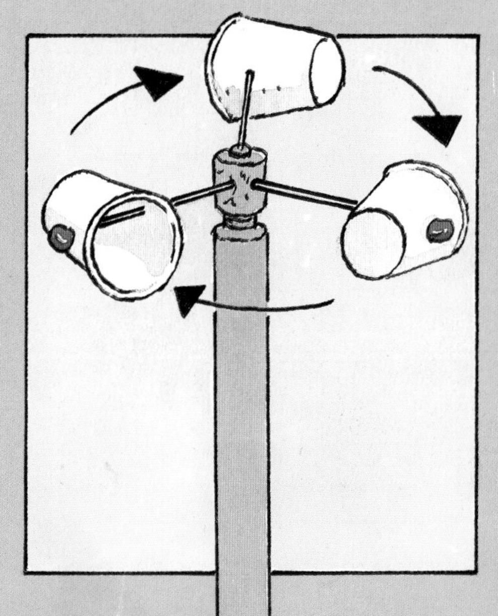

Wind socks and vanes

Which way does the wind blow?

A weather vane turns with the wind. It points in the direction from which the wind blows.

A wind sock billows out in the direction the wind is blowing, or downwind.
This girl has made some wind socks and weather vanes.

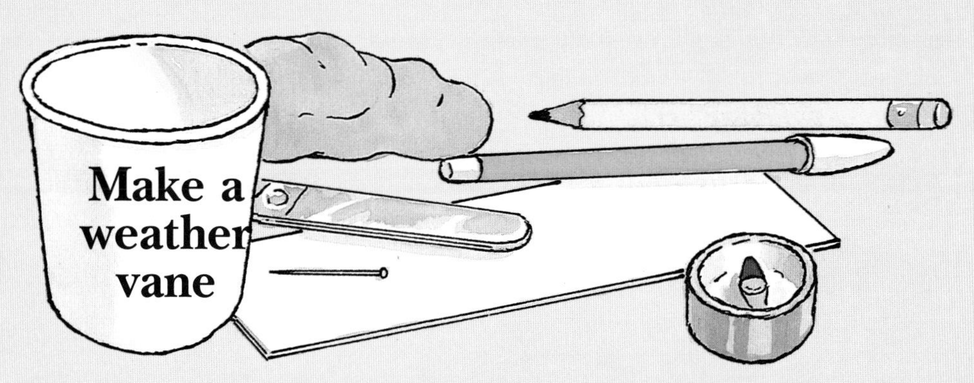

Make a weather vane

Materials: A paper cup, a pencil with an eraser on the end, cardboard, a straight pin, a felt-tipped pen, a compass, and modeling clay.

1. Push the pencil through the bottom of the cup.

2. Make a cardboard arrow with a small slit on its flat end. Cut a smaller cardboard strip and push it into the slit.

3. Push the pin through the arrow into the eraser.

4. Fix your weather vane firmly with modeling clay. Mark North, South, East, and West on the cup.

Highs and lows

A beach ball has a lot of air crowded inside it. This air is at high pressure. When you take out the stopper, the air rushes out to where the pressure is lower.

A barometer measures the air pressure all around us. It can tell us what the weather will be. High pressure means good weather. Low pressure brings wind and rain.

Scissors

Make a barometer

Materials: A balloon, an empty can, a rubber band, a straw, tape, scissors, and cardboard.

1. Cut the neck off the balloon and stretch it over the can.
2. Hold it in place with the rubber band.
3. Tape the straw to the center.

4. Draw a scale on the cardboard.

See how the straw pointer moves as the weather changes.

Make a weather chart

Record the readings that you take with your rain gauge,

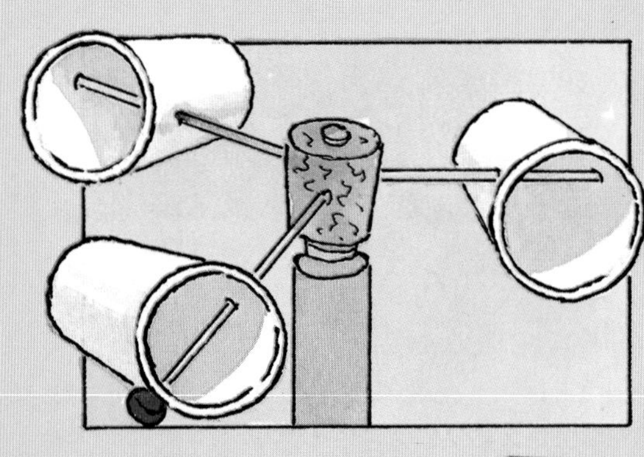

anemometer, weather vane, and barometer on a weather chart.

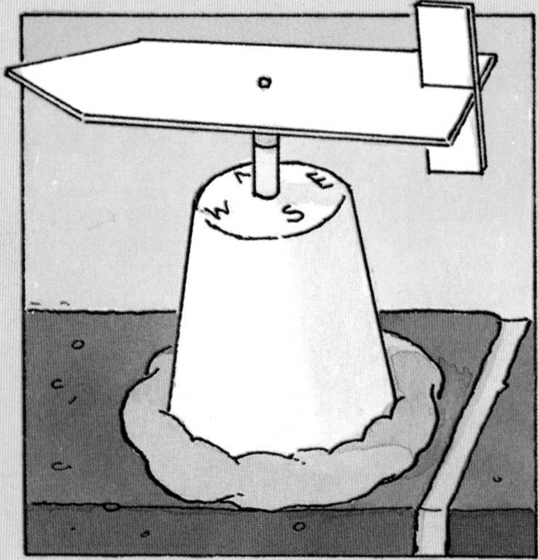

You will also need a thermometer to measure the temperature.

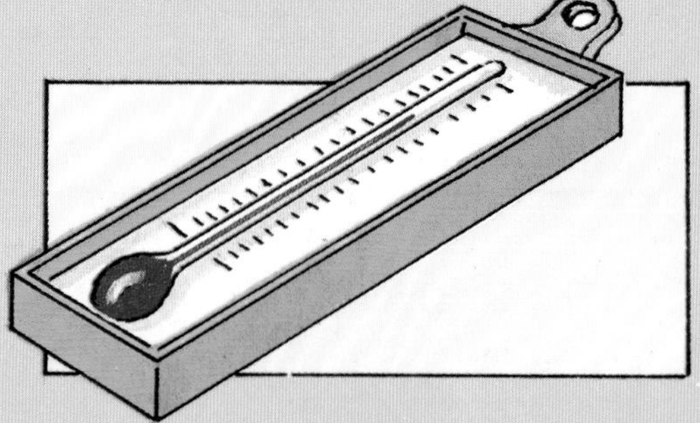

Materials: A large sheet of paper, a ruler, and some colored pencils.

Draw a chart like this one, or make up your own.

WEATHER CHART

Day	Rain	Wind	Wind direction	Air pressure	Temperature	Symbol
Sunday	None	Breeze	West	High	64°F	
Monday	None	Breeze	West	Low	59°F	
Tuesday	1 cm	Strong	North	Low	55°F	
Wednesday						
Thursday						
Friday						
Saturday						

Each day, write in your measurements.

Think about... weather

Satellites orbiting the Earth send us pictures of weather all over the world. This photo shows a hurricane.

At a weather station, workers measure temperature, air pressure, wind direction, and rainfall.

Wet your finger and hold it up in the air. It will feel cold on the side from which the wind is blowing.

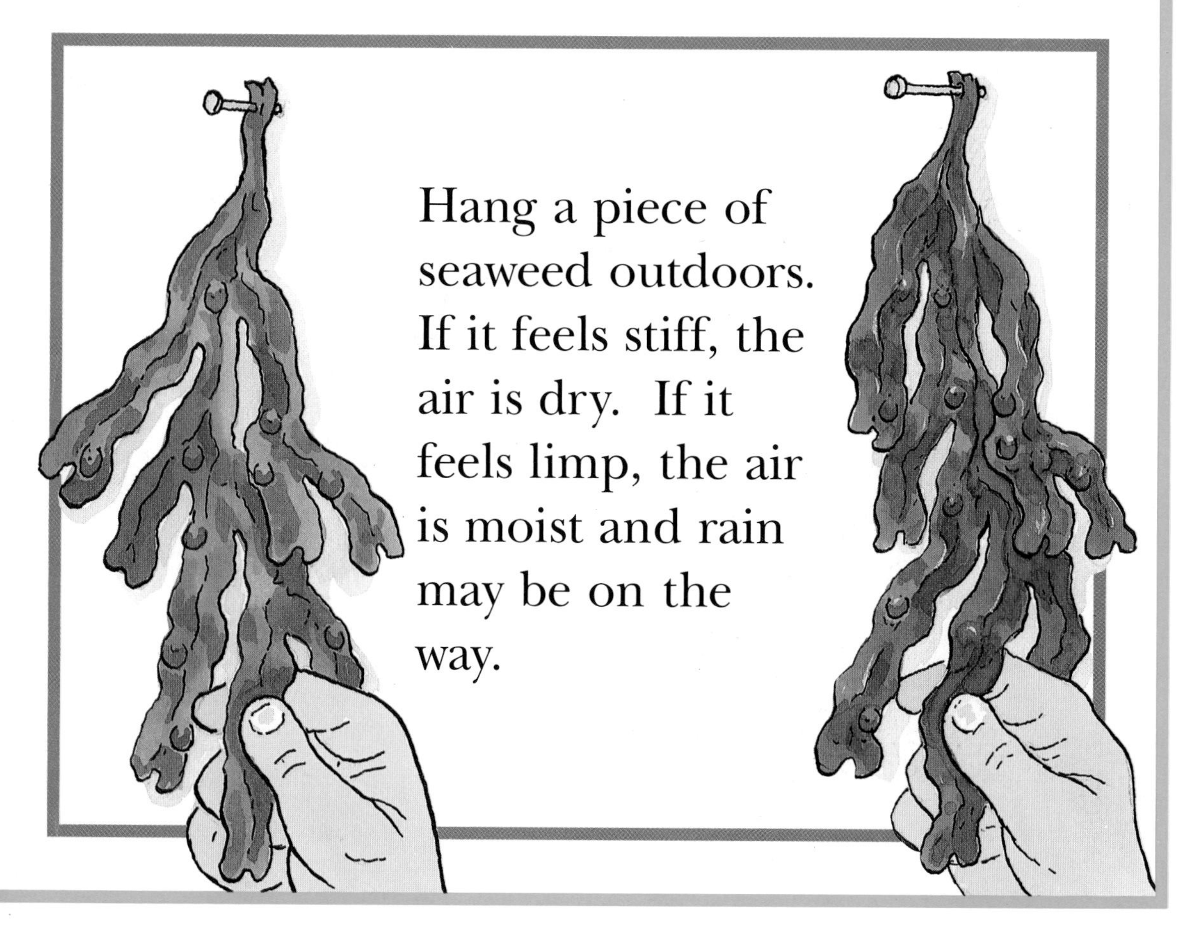

Hang a piece of seaweed outdoors. If it feels stiff, the air is dry. If it feels limp, the air is moist and rain may be on the way.

INDEX